Association of Teachers of Mathematics

Thinking FOR Ourselves

Gold dust!

- Show me an example of...
- What is wrong with the statement? How can you correct it?
- What is the same & what is different about...?
- How can you change...
- Is this always, sometimes or never true? If sometimes, when?
- Convince me that...

avoid "why" ∴ you want to focus on answer not question

Published in 2007 by
Association of Teachers of Mathematics
Unit 7 Prime Industrial Park,
Shaftesbury Street
Derby DE23 8YB

Telephone 01332 346599
Fax 01332 204357
E-mail: **admin@atm.org.uk**

Printed in England

ISBN 978-1-898611-46-2

Copies may be purchased from the above address or **www.atm.org.uk**

Acknowledgments
The authors are grateful to Chris Bills, Liz Bills, John Mason, Malcolm Swan and Anne Watson for ideas which influenced or appeared in the ATM publication, Thinkers.

References
Thinkers ATM (2004)–Chris Bills, Liz Bills, Anne Watson and John Mason

Solving Word Problems in Key Stage 2 PCET (2001)–Jill Mansergh

Zur Fiktivität der 'Stunde Null' im arithmetischen Anfangsunterricht. In Mathematische Unterrichtspraxis. No 4 (1994)–Christoph Selter

■ CONTENTS

∎ INTRODUCTION

This book provides a variety of contexts in which children are encouraged to think for themselves. It is organised into three sections each with a distinct character. The content of the activities in Thinking for Ourselves is intended for pupils working within levels 1 to 5 of the National Curriculum.

Section 1 relates closely to the ATM publication *Thinkers – a collection of activities to promote mathematical thinking* in that some of the section headings are the same. This section encourages children to engage in thinking about mathematical statements in a variety of situations, to respond with their own thoughts and ideas and provide reasons for their responses. The statements are quite often ambiguous or give rise to a number of different answers. The responses of children can be used to provide assessment information about the understanding of particular concepts and, more importantly perhaps, show how powerfully children can think about mathematics.

Section 2 is a collection of longer activities intended to last for at least a lesson. These activities promote all the aspects of Using and Applying Mathematics within interesting contexts.

Section 3 is a collection of pictorial resources. These can also be downloaded from the ATM web-site **www.atmbuy.net/dis019** for use with an interactive white board. The sustainable aspect for this is for children to ask their own questions using a given question as a stimulus. Some examples of open ended questions and more challenging questions are given for the teacher to model with the children. The whole purpose of this section is to provide opportunities for children to ask their own questions.

Supporting the development of thinking skills

This book encourages the development of children's thinking skills. What does this mean and how will the activities in the book help?

In order to develop thinking skills children need to engage in higher order thinking; that is thinking that involves some degree of complexity. Very few of the questions in this book have straightforward answers and many require children to explain their thinking strategies and to offer alternative answers. For instance in the activity *Odd One Out* children are required to make each member of the set the odd one out in turn, giving reasons for their choices at each stage. This is not as simple as it sounds and can be a sophisticated activity which encourages children to think 'outside the box'.

The realisation that there may not be just one answer to a question and the ability to see multiple solutions are higher order thinking skills. Conventional mathematics depends on a set of rules that have been generated by mathematicians. Children can be creative and come up with alternative answers in different situations if they act mathematically and define the rules or the constraints. The activities contained in *Additional Conditions* give a sense of what it feels like to change constraints or rules governing a situation and encourage children to develop situational questions for themselves. This ability to change constraints is fundamental to the 'What if...?' situations with which we expect children to engage. Some longer activities, with the opportunities to apply the skills learned during the shorter questions in Section 1, are included in order to encourage this engagement.

Thinking skills involve a degree of self-regulation, or thinking for oneself. If children are continually prompted towards the next steps they do not develop this valuable way of working which develops their confidence. All the activities in this book are very definitely about children thinking for themselves and not simply being told what to do. Activities that prove too difficult at first should be put to one side. Questions can be left to simmer, perhaps placed on the board and ignored, until children are ready to come back to them.

Thinking skills involve either imposing meaning or finding structure. The success of these activities depends on children finding the meaning or structure for themselves. All pupils should have 'Eureka' moments. The teacher can help children experience these moments by asking appropriate questions to move the thinking on or returning to things when children are ready later.

In many instances the questions ask pupils for reasons or strategies. It is important to identify strategies as this is when metacognition is taking place, that is thinking about thinking or actions. When children engage in this process they make sense for themselves. The process of working on a task generates the act of thinking and the answer is a by-product of the thinking that takes place and the reflection on that thinking.

The activities in *Burying the Bone* have multiple answers and therefore require considerable thought. The process of arriving at an answer is more important than the answers themselves. The questions also allow the teacher to make an assessment of the areas of mathematics with which the children feel confident. Children will only come up with answers related to concepts that are well-embedded. For example in *Lists* children are asked to 'List the set of subtractions with an answer 40'. A child who has been introduced to decimals but is not confident in using them is unlikely to include decimal sets in their answers. Others in the same class will be sufficiently confident to include decimal sets.

Many of the activities focus mainly on oral discussion with some recording being done by the teacher. This method of working allows for mathematical talk to take place. Vygotsky stated, 'The thinking that is generated by the talk between pupils gradually becomes internalised by the individual, so that the group's collective thinking becomes their own.' This reflects the intentions behind the activities in this book in two ways: firstly that children can each learn from the group discussion; and secondly, that these kinds of activity can become mathematical habits so that in every maths lesson they might decide for themselves to sort, to find 'the odd one out', to compare … and so on.

An approach to solving word problems

Solving word problems is difficult for many children. If told that a fisherman catches 12 cod and 6 mackerel and asked how old the fisherman is, many children will give the answer 18. Similar examples are cited in research by Selter (1994)[1]. There are many examples of mathematically able Y5 children replicating this research by giving similarly ridiculous answers to problems. Why is this? Some children have responded that they would not have been given a question which had no answer and that the expectation was for a correct answer. Some tried to justify the answers they had given, for example explaining that the fisherman was given a fish as a birthday present each year.

Perhaps the reason for this is that children are presented with word problems to which they have to work out the answer, long before they are asked to consider framing questions. Often in this type of lesson the extension activity is 'now make up some questions of your own for a friend to solve.' Why should this be the extension? Why not start with the children asking the questions? If the children ask the questions, they are likely to be able to come up with the answer, because in framing the question they have an understanding of exactly what is required.

It is well documented that outside the classroom young children continually ask questions. However, putting them in the classroom situation their rate of asking questions rapidly diminishes. When they do ask a question, a high percentage of them are 'permission' type questions.

[1]Selter, Christoph (1994) *Zur Fiktivität der 'Stunde Null' im arithmetischen Anfangsunterricht. In Mathematische Unterrichtspraxis. No 4.*

The problem solving process

Whilst the intention in this section is to encourage children to pose their own questions, they also need to have a structure that can support them as they work towards the solutions to the questions posed. In solving word problems we go through the process of posing the question in a real life situation, turning this question into a mathematical calculation, solving the calculation and then interpreting this in a real life situation. The final stage is extremely important. Does the answer make sense? Some children may find it useful to use a problem-solving sheet to go through this process. (*see page 38*)

The problem solving process is shown in five sections and needs to be demonstrated by the teacher, with each stage being explained, to ensure children understand exactly what is required.

1. What is the question?
   Frame the question – this might well follow on from work in literacy lessons on asking questions.

2. What is the calculation?
   In this box the children are only required to write in the calculation, not to solve it.
   It is worth reminding them that in the SATs questions 'Show your working you may get a mark' means 'What is the calculation?'

3. Answer to the calculation
   Work out the calculation. Explain that this is not a detailed explanation of how they solved it mentally – just the numerical answer with any jottings is all that is required. This box can also be used for checking answers using the inverse operation.

4. Answer to the problem
   Go back to the question and make sure that this is written in a sentence.

5. Does this answer make sense?
   Think carefully about the answer – is it sensible? It is worth demonstrating an arithmetic error so that the children understand the value of this stage.

SECTION 1—QUESTIONS TO PROMOTE MATHEMATICAL THINKING SKILLS

This section contains a number of short activities each with a different focus but with an underlying rationale. Children should be encouraged to vocalise their thoughts and provide reasons for their observations or conclusions about the varying situations. Differences in answers should be highlighted and explored so that understanding of these differences can be celebrated.

The questions themselves provide many opportunities for assessment of understanding. This is different from the assessment of technique in doing questions. Children who add 32 and 45 correctly may be doing this in different ways and it is the explanation of how they reach the answer mentally that enables assessment of their understanding of addition. Those who say 2 + 5 = 7 and 3 + 4 = 7 so the answer is 77 have a more limited understanding of place value and its relation to addition than children who tell you they partition 32 into 30 and 2, 45 into 40 and 5 and then add the multiples of ten and the units. The second explanation allows a successful evaluation of their understanding of both place value and addition.

The questions in the activities may be used with a whole class as a starting point for talking/partner discussion, for small groups or individuals. This helps the teachers to assess, not only mathematical knowledge, but also children's ability to think mathematically. The questions also provide formative assessment opportunities to help determine next steps for groups or individuals. Each activity contains an example of how to use the questions.

Answers often throw up further questions so it may be useful to use some generic questions, such as those found in the *National Numeracy Strategy Mathematical Vocabulary Book (December 2000)*, to develop children's ideas. It is important that children listen to others and begin to develop the questioning skills that challenge their peers. In using generic questions to follow-up the teacher is modelling the skills that the pupils themselves need to acquire.

Activities in this section are:

- Give me an example of ... and another... and another ...

- Hard and easy

- The same and different

- Odd one out

- Additional conditions. Give me an answer ... then

- Always, sometimes, never true

- Sorting

- Equivalent statements

- Burying the bone

- Lists

- Ordering

- Find the correct solution

- Agree or disagree

- Tell me more

Give me an example of ... and another and another ...

The instruction 'Give me an example of ...' starts with a generalisation and asks children to provide particular examples.

Number

- An even number
- A pair of numbers with a sum of 10
- A pair of numbers with a difference of 2
- An addition with an answer of 6
- A multiple of 2
- A fraction smaller than a half
- A pair of numbers with two factors in common
- A pair of numbers which, when multiplied, give a multiple of 10
- A multiple of 7 which is greater than 100
- Three consecutive numbers with an odd total
- Two fractions with a total of 2
- A prime number
- A five digit number where the hundreds digit is 3
- A fraction that is equivalent to ¾
- A fraction that is equivalent to 20%
- A decimal between 1.3 and 1.4

Shape, space and measures

- A shape that rolls
- A shape with only straight edges
- A shape with four corners
- A pentagon
- A hexagon with just one line of symmetry
- A polygon with a pair of opposite angles equal
- A polygon with an area of 5cm²
- A rectangle with a perimeter of 24 cm
- A shape with parallel sides
- An object with one edge smaller than 20cms
- A container that holds less than a litre
- A small object that is heavy
- A solid with six faces
- A solid with at least four square faces
- A concave shape

Handling data

- A group of objects that have a feature in common
- A diagram or chart you might use to show the number of different pets owned by your class
- A set of 5 numbers with a range of 10
- A set of measures with a median of 5 cm
- A set of numbers that does not have a mode
- An event with a probability of less than a half
- Three numbers with a mean of 7

Consider the instruction, Give me a pair of numbers that differ by 2.
Pause for the children's response, 'and another pair', pause for response, 'and another pair'.

Ask children how they constructed their examples so they are encouraged to consider how to find, for example, a pair of numbers that differ by 2. Responses might vary from 'I just knew it', to 'I thought of a number and added 2 to it', or 'I thought of a number and subtracted 2.'

Older children might, after some thought, give the response '2.1 and 4.1', or '2½ and 4¼'. The reason for the ... *and another, and another* is to allow the children to think more deeply about the problem.

Hard and Easy

Give an example that is really complicated and one that is very simple. Children might then be asked to explain what makes the examples hard or easy.

Number

- A calculation with an answer of 7
- A subtraction calculation
- A pair of numbers that multiply together to give an answer of 64
- A fraction equivalent to ⅔
- A decimal just less than 0.5
- A percentage question involving finding 5%
- A division question
- A word problem
- A prime number greater than 100

Shape, space and measures

- A triangle
- A shape with rotational symmetry order 3
- A shape with line symmetry order 5
- A shape with line symmetry and rotational symmetry order 1
- A rectangle with area 10 cm².
- A shape with a perimeter of 15 cm
- A method for measuring a litre of juice when the litre jug is lost

Handling data

- A set of toys that can be sorted in two different ways
- A list of numbers that can be sorted in two different ways
- A pictogram
- A bar chart showing data about ourselves
- A pie chart
- A set of data with a mean of 5
- A conversion chart
- A distance time graph
- A set of numbers where the probability of choosing a prime number is ½

Consider finding an example of a calculation with an answer of 7.
A hard example for children just getting to grips with the idea of subtraction might be 12 – 9 but it might be 4.9 ÷ 0.7 for those understanding division of decimals. The questions are designed so that children tell the teacher what they think is hard or easy. The teacher gains insight into the level of understanding of the child.

These are useful assessment activities as children may reveal a good deal about what they find difficult and give the teacher an understanding of why they flounder in various situations.

The Same and Different

What is the same and what is different about these pairs? The same and different can be used with objects in the foundation stage and is a good base from which children learn to explain their reasoning.

Number

- 7, 1
- 3, 10
- 2, 5
- 6, 16
- 6, 60
- ½, ¼
- 0.5, ½
- 30%, ³⁄₁₀
- 25%, 70%
- ⅓, ⅖
- ⅓, ²⁄₆
- 3.47, 2.58
- 121, 256

Shape, space and measures

- Square, triangle
- Circle, rectangle
- Line, circle
- Parallelogram, rhombus
- Cube, square
- Square based pyramid, square prism
- Hexagonal prism, pentagonal prism
- Isosceles triangle, equilateral triangle
- Octagon, octagonal prism
- Reflection, rotation
- (2, 3), (3, 2)
- (-1, 2), (-1, -3)
- 30°, 150°
- 70°, 290°
- A litre, 1000ml

Handling data

- Block graph, pictogram
- Carroll Diagram, Venn Diagram
- Bar chart, tally chart
- Sorting, ordering
- Median, mode
- Average, mean
- Bar line chart, line graph
- Likely, unlikely
- Chance, probability

Consider the pair 7 and 1
Children may give a variety of answers. When looking for similarities for example children may suggest they are both odd, they are both less than 10, the numerals are both made up of straight lines, they are both factors of 7. For differences they might come up with 7 is prime and 1 is not prime, 7 is a multiple of 7 and 1 is not, 7 is greater than 5 and 1 is less than 5.

Finding similarities and differences between pairs is a step on the way to asking children to explain their reasoning. There are many different answers to these pairs and these can give an insight into the child's level of understanding of a variety of mathematical concepts.

Odd One Out

Ask children to make each of these items in turn the odd one out of the three, giving their reasons. They should be encouraged to try to make each member of the set the odd one out. The explanation of their reasoning is the most useful aspect of this activity. Here we are building on the previous activity finding things that are the same and different about pairs, but with an extra level of sophistication. This can be extended by asking children to add another item to the list so that their chosen item is no longer the odd one out.

Number

- 3, 4, 9
- 2, 3, 11
- 2 + 4, 3 + 6, 4 + 9
- 8 − 3, 18 − 3, 3 − 8
- 78 − 72, 54 − 37, 23 − 84
- 1, 10, 100
- ½, ¼, ¾
- 2, 91, 89
- ½, 0.5, 50%
- 1:2, ⅓, 33⅓%
- 3 × 7, 13 × 0, 15 × ½
- 25 ÷ 5, 12 ÷ 0, 15 ÷ ½

Shape, space and measures

- Circle, square, triangle
- Cylinder, sphere, cube
- Circle, cylinder, cube
- Left and right, up and down, round and round
- Face, side, corner
- Angle, corner, vertex
- Monday, Wednesday, Saturday
- Metre, minute, ruler
- 60, 360, 100
- Isosceles triangle, equilateral triangle, scalene triangle
- Rhombus, kite, rectangle
- Gallon, metre, gramme

Handling data

- Sort, list, group
- Bar chart, block graph, pictogram
- Vote, survey, questionnaire
- Tally chart, pie chart, bar chart

Red	ℍℍ	
Blue	ℍℍ	I
White	ℍℍ	

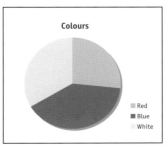

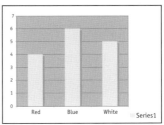

- Impossible, uncertain, certain
- The probability of rolling a 6 on a dice, 0.1̇6̇, ⅙
- Probability of tossing a head on a coin, probability it will rain tomorrow, probability it will be dark when I wake up tomorrow

Consider 3, 4 and 9.
4 could be the odd one out because it is an even number;
3 could be the odd one out because it is prime, or not square,
9 could be the odd one out because the other two are consecutive, or less than five.

This activity leads on from *The same and different* but children are now required to find similarities between two of the elements of the list which are different from the third.

The explanation is the important part with the children being very clear about their reasoning.

Additional Conditions: Give me an example of ... then

Give me an example of... then, provides a starting point to which children give an answer. Then, a further condition is added. Each condition that is added requires children to think more precisely about the examples that they are finding.

Number

- A number ... which is less than 10 ... and more than 5
- An addition sum ... which has an odd number answer ... and the answer is greater than 10
- Two numbers ... with a sum of more than 10 ... and a difference of 2
- A sequence of 4 numbers ... which differ by 2 ... and are all odd
- A group of 6 numbers ... which have a tens digit greater than 3 ... and are multiples of 5
- A multiplication calculation ... whose answer is in the bottom half of a standard 100 square ... whose answer digits differ by 3
- A division calculation ... with a remainder of 2 ... and a divisor of 5
- A percentage calculation ... with an answer of 20 ... and the number you are finding a percentage of is at least 300
- A fraction ... with a denominator which is prime ... and is greater than a half
- A pair of numbers ... with a product of 24 ... and a sum of 11

Shape, space and measures

- A 3D shape ... which can be rolled ... and has at least one flat face
- A 2D shape ... with four sides ... and is not a square
- A 2D shape ... with lines of symmetry ... and 3 corners
- A time on an analogue clock ... that is not an o'clock time ... and the hour hand is between the 3 and the 4
- Show me a container ... which would hold your lunch time drink ... but would not hold a litre of water
- Show me something you could measure with your ruler ... that is less than half the length of the ruler ... that you could also weigh
- Show me a quadrilateral ... which is also a rhombus ... and also a rectangle
- Show me a 3D shape ... which is a prism ... and has 6 faces

Handling data

- 3 playing cards from the pack ... that are red ... that are not picture cards
- A Venn diagram for two sets ... that has 2 members in the intersection ... and 6 members in each set
- 5 numbers with a mean of 6 ... and a median of 6 ... and a mode of 3
- Four numbers with a median of 6 ... that are all odd ... the largest of which is 12
- An event in which a probability of ⅙ might occur ... and a probability of a ⅓ ... and a probability of ½
- An event in which a probability of ¼ might occur ... and a probability of a ⅓ ... and a probability of ½

Consider 'Give me a number ...'

A child might reply 4 or 125 or 3.2 or ¾. Then when asked, 'which is less than 10' they will revise their first suggestion if the number is greater than 10. The suggestion may have to be revised yet again when asked for the number to be more than 5.

Children may need to change their answer to fit the new condition but may be able to give the same answer all the way through.

Always, Sometimes, Never True

Children are invited to say whether a statement is always, sometimes or never true, giving their reasons. They need to find specific examples to illustrate their thinking. This can lead to interesting discussions between the children.

Number

- You can find one more than a number
- If you add a number to an even number the answer is even
- The units digit in a column of a standard hundred square is always the same
- Finding half of something will make it smaller
- Doubling will make a number bigger
- Multiplying an odd number by an odd number gives an odd answer
- Multiplication makes a number bigger
- Dividing by 2 is the same as multiplying by 0.5
- 50% = ½

Shape, space and measures

- A circle has no straight sides
- A triangle has three angles
- A square is bigger than a triangle
- You can roll a cylinder
- A shape with three sides is a triangle
- A regular four sided shape is a square
- If you turn through four right angles you end up facing the same direction.
- The angles of a triangle add up to 180°
- A rhombus has four lines of symmetry
- A quadrilateral with rotational symmetry order 2 is a rectangle
- A polygon with rotational symmetry order 3 also has 3 lines of symmetry

Handling data

- You can sort a set of five objects into two sets
- There will be an overlap when you sort things into two sets
- Information from a tally chart can be put into a bar chart
- A pie chart is clearer than a bar chart
- A Carroll diagram is better than a Venn diagram
- The mean of three numbers is larger than at least one of those numbers
- The mode is always less than the median

Consider 'If you add a number to an even number the answer is even'.
If a child tries this with particular numbers, for example adding 4 to 6, they may believe it is true. They need to try with several numbers, both even and odd, in order to arrive at a generalisation.

Consider 'Finding half of something will make it smaller'. Many children will think that this is always true. Given time, children should be encouraged to look at further examples – the 'What if ...?'. For example: What if we look at fractions? What if we look at negative numbers?

Sorting

The sets of items in each list can be sorted in different ways into two, three or sometimes more sets. There are no right or wrong answers as long as the child can explain how they have sorted the items. There is always an opportunity for the follow up question, 'Can you sort them in a different way?'

Number

- 1, 2, 3, 4, 5, 6, 7, 8
- 3, 6, 8, 13, 15, 16, 18, 20, 21
- 1, 4, 6, 7, 8, 9
- ½, ¼, ¾, ⅓, ⅔, ¹⁄₁₀, ³⁄₁₀, ⁵⁄₁₀, ⁸⁄₁₀
- 16 ÷ 2, 18 ÷ 2, 21 ÷ 2, 15 ÷ 2, 14 ÷ 2, 12 ÷ 2,
- 20 ÷ 5, 16 ÷ 5, 15 ÷ 5, 25 ÷ 5, 80 ÷ 5, 19 ÷ 5, 26 ÷ 5, 65 ÷ 5,
- −1, −3, 0, 6, 19, 51, 53, −17, −83.
- ¾, ½, 0.5, ⅖, 0.75, 0.4, 1.2, 1⅕
- 0.5, 0.45, 0.7, 1.45, 1.5, 1.7, 2.45, 2.5, 2.7
- 10% of 30, 20% of 60, 30% of 60, 60% of 30, 5% of 60, 1% of 300
- 1, 2, 5, 10, 20, 50, 100, 200, 500, 1000, 2000, 5000, 10000, 20000, 50000

Shape, space and measures

- Triangle, square, rectangle, circle
- cone, sphere, triangle, prism, circle
- scalene triangle, isosceles triangle, square, rhombus, parallelogram, equilateral triangle, rectangle
- drawings of rhombus, square, parallelogram, trapezium, rectangle

- drawings of convex and concave shapes

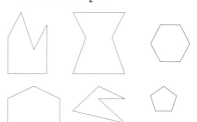

- drawings of 3 × 3 squares with different shadings

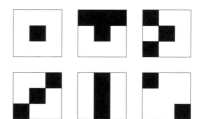

- ruler, tape measure, litre jug, cup, piece of string, 500 gram weight.

Handling data

- Sorting is a data-handling activity so children will engage in data-handling by sorting the sets of items in the first two columns

---

Consider 1, 2, 3, 4, 5, 6, 7, 8
A child might sort them into odd and even; prime and non-prime; multiples of 4, multiples of 3 and the remaining numbers.

Children are making decisions and being asked to justify their choices.

Equivalent Statements

Children are asked to think about another statement which says the same thing as the one given but in a slightly different way or to show that two statements mean the same thing.

Number

- Give me a statement that is equivalent to '4 is less than 9'
- Show that 'a multiple of 2' is the same as an even number
- Show that an 'even multiple of 5' is the same as 'multiples of 10'
- Show that '⅕ of 10' is the same as '½ of 4'
- Show that '5 × 8' is the same as '8 × 5'
- Give me a statement that is equivalent to ' a ¼ of 24 is 6'
- Show that '30% of 200' is the same as '200% of 30'.
- Show that '0.01 × 346' is the same as '0.001 × 3460'

Shape, space and measures

- Show me that a litre is the same as two 500ml
- Show me that 10cm is the same as a decimetre
- Give me a statement that is equivalent to 1000m = 1km
- Give me a statement that is equivalent to 1km = ⅝ mile
- Give me a statement that is the same as 'a square is a rectangle because it has four right angles'
- Give me a statement, that is equivalent to all triangles have three straight sides
- Show me that the four angles of a square sum to the angles in a full turn

Handling data

- Show that 'the mean of 3, 4 and 5' is the same as 'the mean of 2, 3, 4, 5, 6'
- Show that 'the probability of throwing an even number on a dice' is the same as 'the probability of throwing an odd number on a dice'
- Show that 'the range of 2, 3, 4, 6, 7, 7, 8, 9' is the same as the range of ' 2, 5, 6, 9'

Consider 'Give me a statement that is equivalent to 4 is less than 9'.
Children might respond 9 is more than 4, 9 is greater than 4, or 4 is smaller than 9.

Children may answer on very different levels. For example some children may simply work out the mean when asked to show that the mean of 3, 4 and 5 is the same as the mean of 2, 3, 4, 5, and 6. Others may respond by working out the mean of the two sets of numbers whilst others may reason that the mean of 3, 4 and 5 is 4, in the second set of numbers 2 and 6 have been added to the original set so the mean of 2 and 6 is also 4 and therefore the mean of the set of 5 numbers is 4.

Burying the Bone

The answers for each of these activities are focused on a particular type of calculation, but each one can be found in different ways. Children are given a set of conditions and need to come up with a calculation or method which matches the conditions. The children look for different possible ways of finding a solution.

Number

- Give me a question with the answer 5
- Give me an addition with the answer 15
- Give me a subtraction with the answer 3
- Give me a calculation where you add three numbers to give an answer of 24
- Give me a calculation with the answer 6 that involves two additions.
- Give me a calculation with the answer 9 that involves at least one addition and one subtraction.
- Give me a calculation with the answer 25 that involves a multiplication and an addition.

Shape, space and measures

- Give me a method for measuring the length of a table
- Give me a method for measuring a kilogram
- Make me a polyhedron with more than six faces.[2]
- Draw me a polygon with no lines of symmetry.
- Draw me a polygon with a perimeter of 24cm
- Draw me a polygon with an area of 24cm^2

Handling data

- Show me a set of toys that are all smaller than my hand
- Show me ways of sorting this set of toys
- Give me a set of three numbers where the median is 5
- Give me a set of six numbers with a mode of 4
- Give me a set of seven numbers with a range of 6
- Give me a probability question with the answer ⅔

These starting points give children the opportunity to show how much they know. In answering 'Give me a calculation with an answer of 25 that involves a multiplication and an addition', a child might simply give a solution 6 × 4 + 1. However they have the opportunity to explore other possibilities for example 3.1 × 2.5 + 17.25

Those who are very confident can often express their methods in fanciful ways, each one more exaggerated than the last. Children may explore very imaginative possibilities particularly in Shape, space and measures.

---

[2] Use interlocking shapes or ATM Activity Tiles to make polyhedra.

Lists

List or draw a set of objects that satisfies the condition stated. Here children are asked to exemplify a generalisation. It gives them the opportunity for a wide variety of responses at different levels. Encourage children to come up with unique solutions.

Number

- Pairs of numbers that subtract to give 3
- A set of numbers less than 20 but greater then 10
- Pairs of numbers that have 3 as a unit when you add them
- Multiples that end in zero
- Numbers that leave a remainder of 1 when divided by 5
- Trios of numbers that give a total of 25
- A set of fractions that have the same value as 0.5
- The sequences of numbers that contain the number 2
- The set of subtractions with an answer 40
- Factors of 48
- Prime factors of 96
- Multiples of 6 greater than 80
- The set of square numbers greater than 50 and less than 500

Shape, space and measures

- A set of rolling shapes
- The set of shapes with straight sides that can be made by cutting a square into two pieces
- Quadrilaterals with at least one right angle
- Triangles with a right angle
- Coordinates that lie on this diagonal

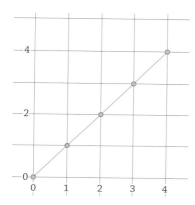

- Units that can be used to measure liquid
- Measures equivalent to 1Kg
- 3D-shapes with more than 6 faces
- The set of polygons with at least two lines of symmetry

Handling data

- All the different ways of sorting a set of animals
- The different ways in which a Carroll Diagram could be constructed to represent the numbers: 5, 6, 7, 10, 11, 12, 15, 16, 17, 20, 21, 22
- Sets of 3 numbers with a range of 8
- Sets of 4 numbers with a median of 5
- Sets of 3 numbers with a mean of 6
- Sets of 3 numbers where the probability of choosing an odd number is ⅔

Consider pairs of numbers with a difference of 3.
A set of responses might include: 3 and 6, 1.2 and 4.2, 100 and 103, −4 and −7, 1 and −2

Ordering

Placing a set of well-chosen objects in order allows children to make connections and see patterns. You may find it helpful to photocopy and cut out the cards on pages 35-36 for children to order. Alternatively ask the children to generate the objects themselves.

Number

- Order these numbers starting with the smallest: 3, 2, 4, 6, 1
- Order these numbers according to how close to 10 they are: 12, 9, 8, 14, 7
- Order these numbers according to how close to 20 they are: 14, 21, 23, 30, 17
- Order these numbers according to how close to 1 they are: ½, ¾, 1¼, 1½, 1¾
- Order these numbers according to how many factors they have: 5, 8, 12, 22, 36

Shape, space and measures

- Order these shapes according to the number of sides: Octagon Quadrilateral, Pentagon, Triangle, Hexagon
- Order these shapes according to the number of faces: Pentagonal prism, Square-based pyramid, Cube, Triangular prism, Tetrahedron
- Order these rectangles according to their perimeter and then re-order them according to their area: 5cm × 6cm, 2cm × 3cm, 10cm × 3cm, 8cm × 4cm, 3cm × 5cm
- Order these measurements using the same units then different units.
  1. 8cm, 3cm, 2m, 3.7cm, 40cm
  2. 6cm, 1m, 2.5m, 47cm, 112cm

Handling data

Order these events according to their likelihood:

- Rolling a 4 on a dice
- Picking a club from a pack of cards
- Tossing a coin and getting a head
- Picking a red card from a pack
- Rolling an even number on a dice
- Order these groups of numbers according to their means

  1, 1, 2, 4, 7

  1, 2, 3, 7, 7

  3, 4, 6, 6, 6

  3, 3, 7, 8, 9

  1, 2, 4, 14, 14
- Re-order according to their medians
- Re-order according to their modes

The questions given are closed and children asked to focus on specific answers.
When children understand what is required, teachers can make this a more open activity by asking them to come up with their own set of numbers to order, according to the given requirement.

Alternatively children can order a given set of numbers and then explain the condition that they have applied.

Extra information for Handling data

Numbers	Mean	Median	Mode
1, 1, 2, 4, 7	3	2	1
1, 2, 3, 7, 7	4	3	7
3, 4, 6, 6, 6	5	6	6
3, 3, 7, 8, 9	6	7	3
1, 2, 4, 14, 14	7	4	14

Find the correct Solution

Several possible answers are given for each question in this section. Children are required to determine the correct solution and give reasons for their choice, then explain how the incorrect solutions have been obtained. You may find it helpful to photocopy and cut out the cards on pages 36 - 37 for children to consider.

Number

- 38 + 12 = ...
  Possible solutions:
  40   10   50   59

- 62 – 27 = ...
  Possible solutions:
  45   35   89   1674

- 38 × 42 = ...
  Possible solutions are:

×	30	8	
40	1200	32	
2	60	16	
			1038

×	30	8	
40	120	32	
2	6	16	
			154

×	30	8	
40	12	320	
2	60	1600	
			1992

- 60 ÷ 2 = ...
  Possible solutions:
  3   30   300   7.5   120

- 5^2 = ...
  Possible solutions:
  3   7   10   25   125

Shape, space and measures

- How many lines of symmetry?

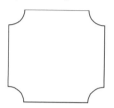

Possible solutions: 0 2 4 5

- Find the perimeter:

3cm

8cm

Possible solutions:
24cm² 22cm 32cm 12cm

- What is the order of rotational symmetry of this shape?

Possible solutions: 0 1 2

Handling data

- Find the probability of rolling an even number on a standard dice.
  Possible solutions:
  ²⁄₆   ⁴⁄₆   ⁶⁄₆   ³⁄₆   ½

- Find the mean of
  3,   8,   2,   12,   9,   2
  Possible solutions:
  6   2   38   5.5   212

The real value of this section is for children to think in a different way and consider in detail possible ways of arriving at solutions.

Agree or Disagree

Children are asked to agree or disagree with a statement and give a convincing argument for their position.

Number

- I always get an answer bigger than 7 if I add a number to 7
- This number is smaller than 100 because it has exactly 2 digits
- Even numbers are always divisible by 2
- Multiples of 4 are always even
- This number is even because it ends with a 4
- The sum of the digits of an even number is always even
- 1 is the smallest prime number

Shape, space and measures

- This shape is a circle because it has no straight sides
- This shape rolls so it cannot be a cube
- This shape does not role so it cannot be a cylinder
- This book is 30cm long because it is exactly the same length as my ruler
- This shape is a quadrilateral because it has more than 3 sides
- I have drawn a cube so a cube must be a 2D shape
- This triangle must have a larger area than this square because I cannot completely cover the triangle with the square

Handling data

- A pie chart is a better way of showing data than a bar chart
- Every set of numbers has exactly 1 mode
- Probability always has to be less than 1
- The mean of 5 numbers is always larger than the median.

Children who disagree should be encouraged to give an example to show the statement is not always the case.

This section links to the ideas of mathematical proof that children will meet in KS3.

Tell me More

Children are given a fact as a starting point and asked to 'Tell me more …' This gives them the opportunity to give simple or more complex answers to further the discussion.

Number

- 8 is a number which only has curved lines
- 2 + 1 = 3
- There are 3 cubes on one side and 10 counters on the other of a balance and the two sides balance
- Numbers ending in 0 are multiples
- 6 × 3 = 18
- I don't like 7 because there are no patterns in the 7 times table,
- 100% is £60
- 75 is ¾ of my number

Shape, space and measures

- This shape will roll
- The classroom is wider than it is long
- I can get inside this shape
- My exercise book is shorter than 30 cm
- If I arrive at 9:05 I'll be on time
- This shape has one line of symmetry
- This triangle is isosceles
- This quadrilateral has two pairs of parallel sides

Handling data

- This chart shows me how the class travelled to school today
- All the counting numbers from 1 to 20 have a place on this diagram
- We can show this information in a pie chart with angles of 60°, 120°, and 180°
- The probability of getting a blue ball is ⅓

Consider 100% is £60. Children work out other amounts from their knowledge of percentages.

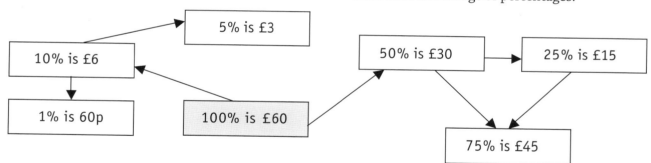

Children work in pairs or small groups with a sheet of A3 paper. Coloured pens might be used to follow different trains of thought, clarifying some of the connections they make.

Some of the starting points given are closed; others are open and give the opportunity for different solutions or statements.

■ SECTION 2 – ACTIVITIES TO PROMOTE MATHEMATICAL THINKING SKILLS

This collection of activities encourages children to think over longer periods of time. The section is organised into the main themes of Number, Shape & space, Measures and Probability. Each of the themes, except Probability, has a set of three activities in ascending order of difficulty, loosely linked to Years 1 & 2, Years 3 & 4 and finally Years 5 & 6.

Probability is only taught at the end of Key Stage 2 and one activity, *In the Bag 2*, is included for this along with some general suggestions for Data Handling.

Each activity is set out as a teaching outline indicating how the activity might proceed, but as with anything in the classroom, different things might happen from those intended. Wherever possible it is important to follow the ideas the children might have allowing them to use their own thinking to guide the process. It is always difficult for teachers to go into the classroom without a clear idea of where the activity might lead. However when teachers have a fixed idea of the outcomes children can be guided too directly towards them minimising the opportunity for children to think for themselves. These activities encourage teachers to try something new and experience the joy of discovering with the children.

There are many opportunities for using and applying mathematics throughout all the activities and the curriculum area addressed is indicated.

Activities in this section are:

■ Number

1 In the Bag 1 – More or Less
2 No Remainders
3 Who Was Sam? – Breaking the code!

■ Shape and space

1 Naming Polygons
2 Perimeters of Polygons
3 Symmetry 1 & 2

■ Measures

1 Comparing Ourselves
2 Giant's Hand
3 Trio Cards

■ Data Handling and Probability

1 In the Bag 2

Number

ACTIVITY 1: IN THE BAG 1 – MORE OR LESS

Begin with some cubes in a bag and add or take cubes from the bag asking children to record how many more or less there are in the bag after each addition or subtraction of cubes.

Ask children to check through their recordings and state how many cubes there might have been in the bag at each stage. (This can be different for different children.)

Next reveal how many cubes were in the bag and ask children to calculate how many cubes were in the bag at each stage using the actual number of cubes.

What is different about their results using an imaginary number of cubes and their results using the actual number of cubes? (The results should show the same difference at each stage of the calculation.)

ACTIVITY 2: NO REMAINDERS

Begin with 12 children at the front of the class and ask them to get into pairs. Stand them in two rows with one of each pair in front of the other. Describe how they are organised in 6 groups of 2. Ask the front row to move forward and now describe 2 groups of 6. Ask what other groupings can be made? Children may say groups of 3 or 4 and should demonstrate by organising and describing the groups.

By continually challenging, the children may try groups of 8 giving the opportunity to group the children in 8s and defining the left-over children that do not make a group of 8 as the remainder. Try other groupings with other remainders.

At this point introduce the notation:
12 put into groups of 8 is 1 group and a remainder of 4.
12 ÷ 8 = 1 R4

Give the pupils sets of cubes and ask them to put them into different groupings, for example groups of six or nine, either with or without remainders and record what they are doing.

Ask what is special about the groups that have no remainders? (They belong in the multiplication table of the divisor.)

ACTIVITY 3: WHO WAS SAM? – BREAKING THE CODE!³

SAM is a square number whose root is a factor of the total number of solutions!

There are many solutions to this problem because there are 6 unknown letters and 10 possible digits that can be used.

For example:

$$\begin{array}{r} \text{WHO} \\ +\ \text{WAS} \\ \hline \text{SAM} \end{array}$$

One solution is

$$\begin{array}{r} 104 \\ +\ 132 \\ \hline 236 \end{array}$$

Set up a display board and write each satisfactory solution to the code on the board as it is found. This will take one lesson to set up and it can be followed up, in registration, or other suitable times, until all solutions have been found and the conditions met to find the solution for SAM!

If the conditions on the picture *(Sam is a square number whose root is a factor of the total number of solutions!)* are too difficult you could use simpler conditions. For example *Sam is the largest number whose digital root is 7.*

The digital root of a number is found by adding the digits together to obtain a single digit. So the digital root of 43 is 7 (4 + 3 = 7), and of 129 is 3 (1 + 2 + 9 = 12 and 1 + 2 = 3).

Some solutions involve a carry digit but the children usually find all the other solutions first. If you tell them there are more someone will eventually think there might be some with carry digits. They should find those that work.

³ This activity comes from a poster produced by the ATM puzzle group, led by Phil Boorman, but is no longer in print.

Shape and Space

ACTIVITY 1 – NAMING POLYGONS

Use a set of ATM Activity Tiles.

Display the polygons. Each one has a pattern on the face.

Ask children to describe any polygons they can see within the patterns.

Outline them in colour as each one is identified on the tiles.

What is the same and what is different about the shapes that have been outlined?

ACTIVITY 2: PERIMETERS OF POLYGONS

Begin by reminding children how perimeters are found.

Give each child a set of ATM Activity Tiles or other regular polygons.

Ask them to use the tiles to help them describe how to find the perimeter of any regular polygon.

Some will find the perimeter of each individual tile but many children will come up with a general rule for a regular polygon. (Take the length of a side and multiply by the number of sides) Children who find the rule can be asked to find the perimeter of a rectangle and explain how it is different from finding the rule for a regular polygon. In this case the sides are different lengths so the rule for finding the perimeter becomes twice the longest side plus twice the shortest side.

ACTIVITY 3: SYMMETRY 1

Use a set of ATM Activity Tiles. If the blank side is shown the number of lines of symmetry and rotational symmetry are the same as the number of sides, but if they are shown the patterned sides this is not true. For instance, the hexagon has no lines of symmetry and rotational symmetry order 3.

Ask children to find the line and rotational symmetries for each polygon.

ACTIVITY 3: SYMMETRY 2

Throw three different coloured dice and add 2 to the first number to find the number of sides of the polygon. Use the other two numbers to give the line and rotational symmetry of the polygon. If the three dice show 2, 4, 1 consider a regular polygon with 4 sides, (2 + 2), line symmetry of order 4 and no rotational symmetry of order 1 (no rotational symmetry).

Can the children design a polygon that fits the conditions described by the dice?

What combinations of throws are possible and which are impossible?

Measures

ACTIVITY 1: COMPARING OURSELVES

Ask groups of children to get into line according to their heights and record the order.

Does this mean that if they order themselves using shoe sizes they will be in the same order as the heights? What about hand spans or..? What can they discover?

ACTIVITY 2: GIANT'S HAND

Use the photocopier to make an enlarged picture of your hand by placing it flat against the plate.

This can be used to tell the children that your house had been broken into and the police have found a handprint. It is obvious that it is a large person and the police would like the children's help in finding how tall the person is and anything else they can find out about his/her dimensions.
A variety of pupils of different ages have trialled this activity. The younger children generally take a more practical approach, for example trying to match the handprint to a large person. Older children often start to take measurements and work out averages for hand size and height and so on.

ACTIVITY 3: TRIO CARDS

The main focus of this activity is on making relationships so that children can make links across a range of measures and note any similarities and differences between them. Give each group of children a set of measures such as litres, metres…. These can be metric and/or if you wish to make it a little more difficult, imperial. Ask the groups to choose 3 related measures for their set and to put them together as a trio on a card. They write down all the facts they know about the three measures and the relationship between them in poster format.

Children then look at the posters that have been constructed by each group and discuss the similarities and differences between the measures that are displayed.

FOR EXAMPLE LITRES, MILLILITRES AND CENTILITRES

There are 1000 millilitres in a litre.

There are 10 millilitres in a centilitre.

There are 100 centilitres in a litre.

A millilitre is a thousandth of a litre and a tenth of a centilitre.

A centilitre is a hundredth of a litre.

The litre is the largest measurement of capacity in the set.

The millilitre is the smallest measurement of capacity in the set.

An example of a similarity between posters that the stem 'milli' occurs when the equivalence is 1000.
An example of a difference is that time is in base 60.

Litres

Millilitres

Centilitres

Data and Probability

Opportunities for activities that involve working with data can be found in various different ways such as:

- Cross-curricular opportunities with links to science, geography, history and other subjects.

- They can also be found from work using measures - see activities 1 and 2 in the previous section.

- The pupils often generate hypotheses themselves. For instance they might state:
  - It always rains on Thursday
  - Everybody likes chocolate
  - Girls don't like to do sports activities
  - Boys are taller than girls.

Each of these hypotheses provides opportunities for data collection, representation and interpretation. Using the children's ideas validates them and also shows how data handling is used in real life.

ACTIVITY: IN THE BAG 2 – WHAT COLOURS ARE IN THE BAG?

Use a bag containing either 10 or 20 cubes in three different colours. One option is to have a lot of one colour, a small number of another colour and the rest of a third colour. Do not tell children how many cubes are in the bag.

Go round the class asking about half of the children to take out and replace a cube. At the end of the first round ask what they can tell you about the cubes in the bag. For instance, they might say, 'There are three colours' but when asked if they are they sure they may think that there could be other colours that have not been pulled out. They might say there are more of certain colours than others. Record all the statements as they are made.

Play another round but this time ask the children to record what happens in a table. To keep any fractions simple use a limited number of children, for example with 10 cubes ask 20 children. When the round is completed ask them to use their information to tell you again about the cubes in the bag. Did they find out anything different from the first set of information where they did not record in table form?

Ask children to write the probabilities of picking a cube of each colour out of the bag based on their recording. Try another round and again record the answers. Is this the same or different from the previous recorded round? Do any conclusions from their previous recorded round need to be changed? What if the two results are added together? What does that do to the conclusions?

Now tell children the actual number of cubes in the bag and the number of different colours it contains. From their previous samplings, can they predict how many of each colour there will be? Finally reveal all! Compare the results from the experiment with the actual number of cubes in the bag. It is a continual surprise how close the theoretical and experimental probabilities are.

SECTION 3 - SOLVING WORD PROBLEMS – CHILDREN ASKING QUESTIONS

This section contains pictures and information, which can be used to encourage the children to ask and answer their own mathematical questions. The contexts give scope for many cross-curricular questions which raise other issues for discussion. The pictures can be downloaded from the ATM website **www.atmbuy.net/dis019** for use on an interactive whiteboard.

In the introductory lesson, the teacher might start by asking a series of closed and open questions. This can be followed up by children working with a partner, asking each other questions and then, sharing some with the whole class. Initially, some children experience difficulty in framing questions, so plenty of discussion at this stage is essential. Evaluate their first attempts and, if the questions are all very simple, you may wish to use one of the possible questions suggested in the activity to model more thought provoking questions for the children. Always encourage variety and complexity and encourage children to review and amend their own questions in light of the discussion. This gives opportunity for strong links with literacy.

Some questions are included which may be used by the teacher to model more challenging questions. Discussion around 'What makes this a more challenging question?' would be valuable. The ideal is for children to come up with the questions themselves based on the premise that if a child can ask a mathematical question then they know how to answer it. Over time a bank of questions can be built up and used in future lessons.

The numbers on the posters are deliberately simple so that children concentrate on the problem solving process and do not get bogged down in the arithmetic. For some children this is not a good approach as they are motivated by the challenge of more difficult numbers. It is helpful to replace the given information with more challenging numbers for these children. They will still be working on the problem solving objective along with the rest of the class, but with the extra challenge. For example, children could be asked to add on VAT at 17.5%.

Activities in this section are:

- Riding stable

- Pantomime

- Swimming Pool

- Miniature Railway

- Go Kart Racing

Possible extensions:

- Give me a question where you need to use addition to find the answer.

- Give me a question where I would have to look at more than one of the posters to find an answer

- Plan an activity day using all the information sheets. How much would you spend?

- If you were to have a week's holiday, draw up a timetable of your activities and calculate how much money you would need. You may decide to make up some more information sheets to fill your week.

Riding Stable

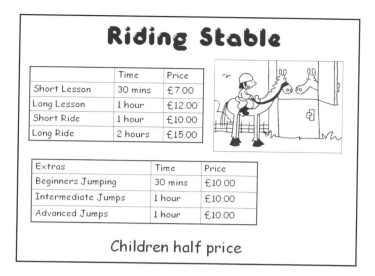

STRAIGHTFORWARD QUESTIONS

- What is the cost of a long lesson? *(£12)*

- How long does a short ride last? *(1 hour)*

- What would be the cost of a short lesson followed by beginners jumping? *(£17)*

- How long would Sam be at the riding stable if he had a long ride followed by an advanced jumps session? *(3 hours)*

- If you spent two hours at the stables what would be the cheapest / most expensive way to spend your time? *(Multiple answers e.g. long ride for the cheapest, 4 lots of beginners jumping for the most expensive. Children may decide that activities cannot be repeated.)*

- Amy had £30 to spend at the stable. How could she make this last the longest amount of time? *(Multiple answers e.g. 2 Long Rides)*

CHALLENGING QUESTIONS

- What are the possible ways of occupying 3 hours? *(Multiple answers e.g. Long Ride and Intermediate Jumps)*

- What activities would you consider if you had £30 to spend at the stable? *(Multiple answers e.g. Long Ride and Advanced Jumps)*

MORE CHALLENGING QUESTIONS

- John says 'It costs exactly the same for the beginners jumping and the advanced jumping'. Do you agree with his statement?

- Which is better value the short lesson or the long lesson? Give your reason.

- If the riding stable have a special offer for children allowing them to participate in events all day for a special price of £25 would this be good value?

POINTS TO CONSIDER

- This might lead to an interesting discussion

- Encourage the children to compare times for equivalent money, or compare the cost for equivalent times.

- Some children will maintain that the short lesson is better value because it is cheaper.

- This provides a starting point for a wide variety of discussions. It would be useful for the children to work in pairs to come up with a reasoned argument for or against.

Pantomime Cinderella

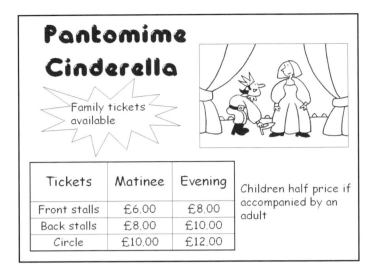

STRAIGHTFORWARD QUESTIONS

- How much would it cost for a back stalls seat on Friday evening? *(£10)*

- What is the cost for a mother and her 7 year old son to watch Cinderella from the circle on Saturday afternoon? *(£15)*

- Jo's auntie takes Jo and her two sisters to see Cinderella on Wednesday afternoon. They sit in the front stalls. Jo's auntie has £20 to spend on the outing. Could they have afforded to go in the back stalls or the circle? Could they have afforded to go in the evening?
  (Back stalls for Matinee and Front stalls in the evening.)

CHALLENGING QUESTIONS

- Five children and an adult want to go and see Cinderella. They have a budget of £30. What possibilities are open to them? *(Multiple answers e.g. Matinee - Front stalls)*

- A Brownie pack is arranging a visit to the pantomime. There are 15 Brownies with three adult helpers. How much would it cost for them all to go? *(Multiple answers e.g. the cheapest would be all in the Front stalls for a matinee, £63)*

MORE CHALLENGING QUESTIONS

- Is it reasonable to have different prices for different seats? Why do you think theatres do this?

- Why do you think the children can only have a half price seat if, they are accompanied by an adult?

- What do you think the price should be for a family ticket? Give a justification for your answer.

POINTS TO CONSIDER

- These are the types of questions where every child can have an answer, because they are opinion-based.

- It might be useful to have a 'no-hands-up' policy in order to encourage all children to be ready to answer.

- There is a lot of mathematical reasoning attached to this question. It would be an ideal question to discuss with a partner.

Swimming Pool

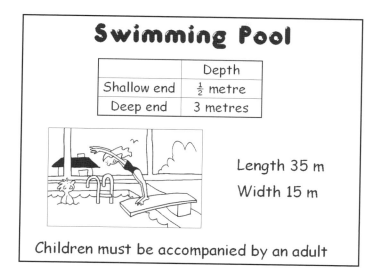

STRAIGHTFORWARD QUESTIONS

- Ben managed to swim 2 widths of the pool. How far did he swim? *(30 m)*
- Lily swam 4 lengths – how far is this? *(140 m)*
- Amy took 40 minutes to swim 20 lengths – on average how long did each length take? *(2 minutes)*

CHALLENGING QUESTIONS

- Where would the water come up to on you in the shallow end?
- Do you think you would be able to reach the bottom in the deep end?
- How many people do you think can safely swim in a swimming pool of this size?
- At busy times the swimming pool will allow people in for a fixed length of time. How long do you think this should be?

MORE CHALLENGING QUESTIONS

- Sean had a new pair of shorts for his birthday. Would he be able to paddle in the shallow end without getting his shorts wet?

- Ann can swim 3 widths – do you think she would be able to swim a length?

- Adam can swim all the way round the edge of the pool – how far is this?

POINTS TO CONSIDER

- Obviously here Sean's height is very important. Encourage children to think what other information they might need to answer the question.

- Children invariably come up with answers such as 'She gets a rest each time she swims a width so not necessarily', or 'I can swim loads of widths, but I wouldn't go into the deep end'.

- Children may be extremely logical and say that it is not 100 metres because Adam does not swim all the way into the corners so it is a little less than 100 metres.

Miniature Railway

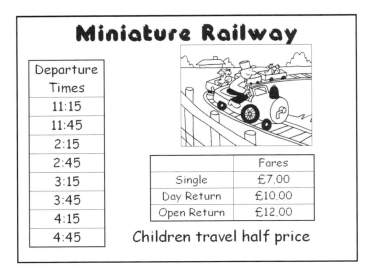

Miniature Railway

Departure Times
11:15
11:45
2:15
2:45
3:15
3:45
4:15
4:45

	Fares
Single	£7.00
Day Return	£10.00
Open Return	£12.00

Children travel half price

STRAIGHTFORWARD QUESTIONS

- How much would it cost for 2 adults and two children to purchase a day return ticket? *(£30)*
- Natalie aged 10 and her 8 year old sister and Anna catch the 3:15 having bought two single tickets. How much did they have to pay? *(£7)*

CHALLENGING QUESTIONS

Extra information

The train runs along a circular route within a safari park and is 6 miles long. It runs between the station by the large fairground rides and the station by the house which people can look around.

- What do you think the train timetable would look like if you were making the return trip?
- The departure times are from the fairground station. I need to get to the house station for 3:00. Which train do I need to catch?
  (Multiple answers, depending on where on the circular track the stations are.)

MORE CHALLENGING QUESTIONS

- How long do you think the train journey is one way?

- Why are the departure times not evenly spaced?

- What are the advantages of purchasing a return ticket?

POINTS TO CONSIDER

- What does long mean? - A time? A distance? Are there some questions that the children need to ask before attempting to come up with a solution?

- This question encourages children to speculate. Encourage them to come up with more than one reason.

- There are different types of returns. Make sure each type is considered.

Go-Kart Racing

Go-Kart Racing

Races	Start Time	Entrance Fee
Under 12s	2:00	£10.00
Under 16s	3:00	£15.00
16 and over	4:00	£20.00

	Prizes
Winner	£50.00
Second Place	£30.00
Third Place	£20.00

STRAIGHTFORWARD QUESTIONS

- 7 year old Adam comes third in a race. How much money does he make out of the race? *(£10)* Discussion point – prize money minus entrance fee.

- 16 year old Harriet comes third in a race – how much money did she make out of the race? *(Nothing)*

- Two brothers aged 10 and 12 enter races. How much does their father pay in entrance fees? *(£25)*

CHALLENGING QUESTIONS

- There were 24 participants in the 16 and over race. How much profit did the organisers make? *(£380)*

- 60 children entered for the under 12s race. This was too many children to race at one time. How might the organisers have arranged this so that there were a maximum of 20 cars on the track at any one time, but they only gave away prizes for the overall first second and third places? *(Multiple answers e.g. 3 heats with 20 cars and the first 6 in each race given places in the final.)*

MORE CHALLENGING QUESTIONS

- How many people do the organisers need to enter the under 12s race to be sure of breaking even?

- Why do you think there might be different entrance prices depending on the age of the entrants?

POINTS TO CONSIDER

- This could be an ideal situation for changing the prices for more able pupils. If £10.00 were changed to £9.95 rounding strategies could be used.

- Encourage the children to speculate, giving more than one reason.

6	14	30	1¾	36
4	12	23	1½	22
3	9	21	1¼	12
2	8	17	¾	8
1	7	14	½	5

octagon	pentagonal prism	5cm × 3cm (rectangle)	rolling an even number on a dice	1, 2, 3, 14, 15
hexagon	tetrahedron	8cm × 4cm (rectangle)	picking a red card from a pack	3, 3, 7, 8, 9
pentagon	cube	3cm × 2cm (rectangle)	tossing a coin and getting a head	3, 4, 6, 6, 6
quadrilateral	square based pyramid	10cm × 3cm (rectangle)	picking a club from a pack of cards	1, 2, 3, 7, 7
triangle	triangular prism	6cm × 5cm (rectangle)	rolling a 4 on a dice	1, 1, 2, 4, 7

1674	59	50	4010	
	89	35	45	
		300	30	
120	7.5			3
125	25	10	7	3

Table (1992):

X	30	8
40	12	320
2	60	1600
		1992

Table (154):

X	30	8
40	120	32
2	6	16
		154

Table (1038):

X	30	8
40	1200	32
2	60	16
		1038

   ATM

5	12cm		1/2	212
4	32cm	2	3/6	5.5
2	22cm	1	6/6	38
0	24cm²	0	4/6	2
			2/6	6

The Problem Solving Process

Problem

Mathematical calculation

Mathematical solution

Answer to the problem

Does this answer make sense?

 Association of Teachers of Mathematics

ATM